B-29 Superfortress

in action

By Larry Davis

Color by Don Greer

Illustrated by Joe Sewell

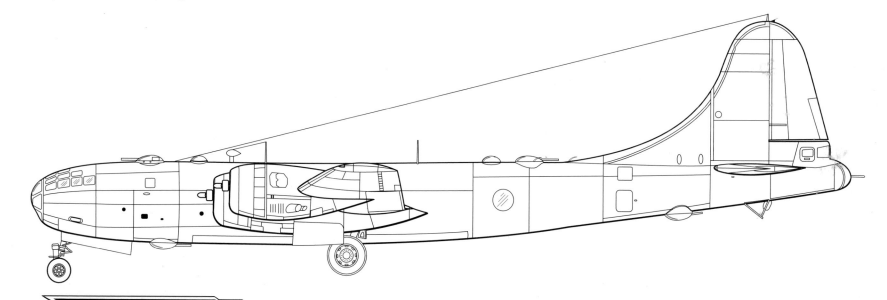

Aircraft Number 165

squadron/signal publications

DAUNTLESS DOTTY, piloted by MAJ Robert Morgan, was the first 21st Bomber Command raid on Tokyo on 24 November 1944. Robert Morgan piloted MEMPHIS BELLE, the first crew to finish twenty-five missions in Europe.

DEDICATION

To the valiant crews of "ENOLA GAY" and "BOCKSCAR", who had to fight first the Japanese during World War 2, and then the revisionist historians of the 1990's. Congratulations on a JOB WELL DONE!

LIST OF CONTRIBUTORS

Peter Bowers
Boeing Aircraft Co.
Joe Bruch
Bob Esposito
Jeff Ethell
Norris Graser
Tom Ivie
James F. Lansdale
Dave Lucabaugh
Ernie McDowell
David Menard
Robert Mikesh
Hans-Heiri Stapfer
Larry Sutherland
USAAF
US Air Force Museum
Nick Williams

ISBN 0-89747-370-1

If you have any photographs of aircraft, armor, soldiers or ships of any nation, particularly wartime snapshots, why not share them with us and help make Squadron/Signal's books all the more interesting and complete in the future. Any photograph sent to us will be copied and the original returned. The donor will be fully credited for any photos used. Please send them to:

Squadron/Signal Publications, Inc.
1115 Crowley Drive
Carrollton, TX 75011-5010

Если у вас есть фотографии самолётов, вооружения, солдат или кораблей любой страны, особенно, снимки времён войны, поделитесь с нами и помогите сделать новые книги издательства Эскадрон/Сигнал ещё интереснее. Мы переснимем ваши фотографии и вернём оригиналы. Имена приславших снимки будут сопровождать все опубликованные фотографии. Пожалуйста, присылайте фотографии по адресу:

Squadron/Signal Publications, Inc.
1115 Crowley Drive
Carrollton, TX 75011-5010

軍用機、装甲車両、兵士、軍艦などの写真を所持しておられる方はいらっしゃいませんか？どの国のものでも結構です。作戦中に撮影されたものが特に良いのです。Squadron/Signal社の出版する刊行物において、このような写真は内容を一層充実し、興味深くすることができます。当方にお送り頂いた写真は、複写の後お返しいたします。出版物中に写真を使用した場合は、必ず提供者のお名前を明記させて頂きます。お写真は下記にご送付ください。

Squadron/Signal Publications, Inc.
1115 Crowley Drive
Carrollton, TX 75011-5010

The end of the Japanese Empire is in sight as B-29s from the 29th Bomb Group taxi into take-off position at North Field, Guam during April of 1945. (Dave Lucabaugh)

INTRODUCTION

As early as 1934 the U.S. Army had envisioned an aircraft that was capable of carrying a 2,000 pound payload, with a range in excess of 5300 miles, which was far beyond current technology! However, by 1940 the Army issued design requirements that had evolved into specifications that called for a 'super bomber' — capable of a 400+ mph top speed, a 2,000 pound payload, and a range in excess of 5300 miles! All this in 1940, when the small, nimble fighters could not top 400 mph; and long range bombers barely had ranges of 2500 miles. What manufacturer could possibly hope to meet these requirements? Four tried, two succeeded.

Boeing, Consolidated, Douglas, and Lockheed all put forth design proposals based on these requirements. Lockheed's **XB-30** (Model 51-81-01) was based on an armed version of what would eventually become the very successful Constellation airliner. However, as a 'super

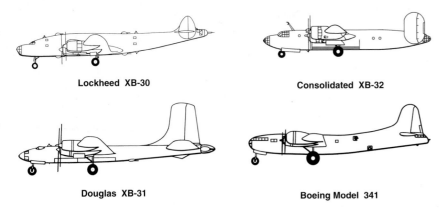

Lockheed XB-30

Consolidated XB-32

Douglas XB-31

Boeing Model 341

bomber' the Lockheed design never went beyond the proposal stage, although a scale model was built. Neither did the Douglas **XB-31** (Model 332F). The Boeing offering, the **XB-29** (Model 341) was accepted and went into production. The Consolidated **XB-32** (Model 33) ultimately went into limited production (116 built) as the B-32 Dominator. But Army Air Force wanted the Consolidated B-32 only as a backup in case the Boeing B-29 program failed.

Boeing, however, had a leg up on its competitors. In the early 1930s Boeing was already at work on the problems of long range aircraft. As part of Boeing's continuing study of long range bombers, the **model 334** design study was a four engine bomber with a tricycle landing gear and a pressurized cabin. The initial model 334 design featured a twin vertical tail and a wing planform similar to that on the successful Boeing model 314 Clipper flying boat. Engines in the model 334 proposal were to be a radically new flat, liquid cooled Wright design, so thin that they would be completely enclosed within the wing structure. However, in July 1939 Boeing unveiled a completely revised design, the **model 334A**. It is from this design study that the B-29 genealogy can be traced.

The model 334A had a circular fuselage with an all-glass nose, a large single vertical tail assembly, and was powered by four Pratt & Whitney R-2800 radial engines. It also featured a high aspect ratio wing, very long and thin — not unlike the 'Davis Wing' used on the B-24. By late 1939, Boeing's Ed Wells had developed the proposal still further, resulting in the

model 341. The model 341, which would be the design submitted to the Air Corps, had a very streamlined, circular fuselage with a single tail. This was mated to a huge 124 foot 7 inch model 115 high aspect ratio wing, which was capable of supporting twice as much weight per square foot of area than any previous bomber wing.

RAF combat experience in the air war over Europe dictated new demands of the new bomber design; Army Air Corps wanted self-sealing fuel tanks and greatly increased defensive firepower. These additions would add some 5,500 pounds to the weight of the model 341, which in turn would require an additional 3,600 pounds of fuel to maintain the bombers range. The model 341 simply was too small to meet the new demands. On 11 May 1940, Boeing unveiled the **model 345**. It would become the world's first 'super bomber'. The new bomber design was to be so superior to its predecessor, the B-17 FLYING FORTRESS, that Boeing simply had to call the new airplane the SUPERFORTRESS.

Chief Aerodynamicist George Shairer developed a new model 117 high aspect ratio wing that could lift almost ninety times its weight! On 24 August 1940, The Army Air Corps placed an order with Boeing to build two prototypes based on the model 345, under the Army designation **XB-29**. A third prototype was ordered in December. The XB-29 mock-up was built and approved in the early spring of 1941. On 17 May 1941, without even one prototype to inspect and approve, Air Corps placed an order for fourteen **YB-29** service test aircraft, and two hundred fifty **B-29** production aircraft, now officially known as the SUPERFORTRESS. This order was doubled in January 1942, just weeks after the Japanese attacked Pearl Harbor, when the War Department realized they would need far more of the new long range bombers to defeat the Japanese.

On 21 September 1942, the **XB-29-BO** was ready. The XB-29 was the first bomber aircraft to be fully pressurized, the first with remote control turrets connected to a central fire control system, the first to be powered by the brand new Wright R-3350 engine, and the world's heaviest production airplane to date. Powered by the newly developed 2200 hp, 28 cylinder Wright R-3350-13 Cyclone, twin row radial engine, the XB-29 would have a top speed of 368

Boeing Design Evolution

Model 334

Model 334A

Model 345

Model 345(XB-29)

mph, a service ceiling of 32,100', and a range in excess of 5800 miles. The XB-29 was 98 feet 2 inches in length, 27 feet 9 inches to the top of the vertical tail, and had a wingspan of 141 feet 3 inches, with a combat ready weight at a hefty 105,000 pounds.

A number of different defensive armament combinations were tried on the XB-29, including manned turrets like those on the B-17 Flying Fortress, and the ERCO remote gun blisters sim-

Rollout of the XB-29 came during early September of 1942. The tail bumper, small tail gunners windows, and larger rudder were found only on the XB-29s (Peter Bowers)

ilar to those found on the Navy PB4Y-2 Privateer. Boeing and the Army finally settled on four Sperry remote control turrets, two of which were mounted on the upper fuselage and two more being mounted on the lower fuselage, each carrying a pair of Browning M2 .50 caliber machine guns. Teardrop shaped sighting blisters on the upper and lower rear fuselage controlled the turrets. In the tail was mounted another pair of .50s coupled with a single 20mm cannon. This three gun tail turret was controlled by a tail gunner, who sat in his own pressurized compartment directly under the rudder. The sighting blisters used periscopic gun sights.

All manned positions were fully pressurized. The bomb bay was not, nor was the aft fuselage between the aft turret gunners station and the tail gunners compartment. A 40 foot long pressurized tunnel that ran through the upper part of the bomb bays connected the forward cabin to the aft gunner's compartment. The XB-29 had two bomb bays, with an intervalometer that regulated the bomb-dropping sequence to keep the airplane's center of gravity balanced. At 3:40pm on 21 September 1942, Boeing Test Pilot Edmund "Eddie" Allen lifted the Olive Drab over Neutral Gray XB-29, serial 41-002, off the runway at Boeing Field for the first

The aft fuselage of the second XB-29 with teardrop shaped sighting blisters, but still without turrets. The tail 'bumper' has now been changed to a retractable tail 'skid'. Test pilot Eddie Allen was killed testing this aircraft. (Peter Bowers)

The 3rd XB-29 on the ramp at Boeing Field in of June 1943, showing the teardrop-shaped astrodome above the canopy. The three blade propellers were common to all three XB-29s. (Peter Bowers)

time. He landed the big bird forty five minutes later. His only comment (with a big grin), "She flies!"

The XB-29 flew extremely well, with the huge airplane initially performing almost flawlessly during all phases of the initial testing, which was especially surprising considering all the technological innovations incorporated in the new bomber. But problems soon began to surface — engine problems. Problems that would persist throughout the life of the B-29 series. Persistent overheating and frequent engine fires became common occurrences as flight testing went above 25,000 feet. On 28 December two engines failed; a week later one of the replacement engines failed. The second prototype, #41-003 made its first flight on 30 December, and had its first engine problem when one of the big Wrights caught fire. These problems continued throughout January and February of 1943, finally culminating in disaster on 18 February, when an inflight engine fire burned through the main wing spar of #2 prototype, taking the life of Eddie Allen, his test crew and a half dozen fire fighters on the ground in the ensuing crash and bringing the test program temporarily to a standstill.

However, to ensure the survival of their long range bomber the Air Corps took over management of the entire project. Reorganized under the command of Brigadier General Kenneth Wolfe and renamed B-29 Special Project, this radical approach worked and the SUPER-FORTRESS program pushed forward.

On 15 April 1943 the first of the fourteen YB-29 service test aircraft came off the assembly line at Boeing's newly built plant in Wichita, Kansas. The YB-29s were virtually identical to the XB-29 except for turrets and the fire control system. Following quickly on the heels of the YB-29, were the first of the production models, designated simply **B-29**.

Special scaffolding had to be built to service the 29'7" tall XB-29 tail assembly. At this time the #1 XB-29, 41-002, had Sperry remote gun turrets. (Peter Bowers)

Many different armament variations were developed and tested for the B-29. This Superfortress has manned B-17 upper turrets, with manned ball turrets under the fuselage, and remote .50 caliber ball turrets on the nose, with hand-held single .50s in each of the waist gun positions. (Peter Bowers)

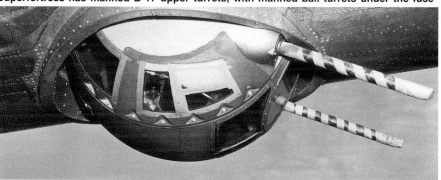

(Above/Below)The lower manned ball and upper turrets carried a pair of .50 caliber machine guns, while both waist positions mounted single .50 caliber machine guns. (Peter Bowers)

(Above / Below) The seventh YB-29 was experimentally fitted with an ERCO ball turret in the nose, and twin gun turrets on the fuselage sides that had originally been developed for the Convair PB4Y-2 Privateer and B-32A Dominator. (Peter Bowers)

7

(Above) The first XB-29 on the portable scales at Boeing during June of 1943. The General Electric remote gun turrets have been installed, as well as new engine cowlings for the Wright R3350-13 engines. (Peter Bowers)

(Right) The sixth YB-29 landing at Boeing Field. The huge Fowler Flaps, which roll back and down, creating a much larger wing surface area without compromising the thin chord of the model 117 wing. (Peter Bowers)

(Below) The first YB-29 was modified by General Motors with the installation of Allison V-3420-11, twenty-four cylinder liquid cooled engines and re-designated the XB-39-GM. A V-3420 engine was two 12 cylinder V-1710 engines mated together. (David Menard)

Development

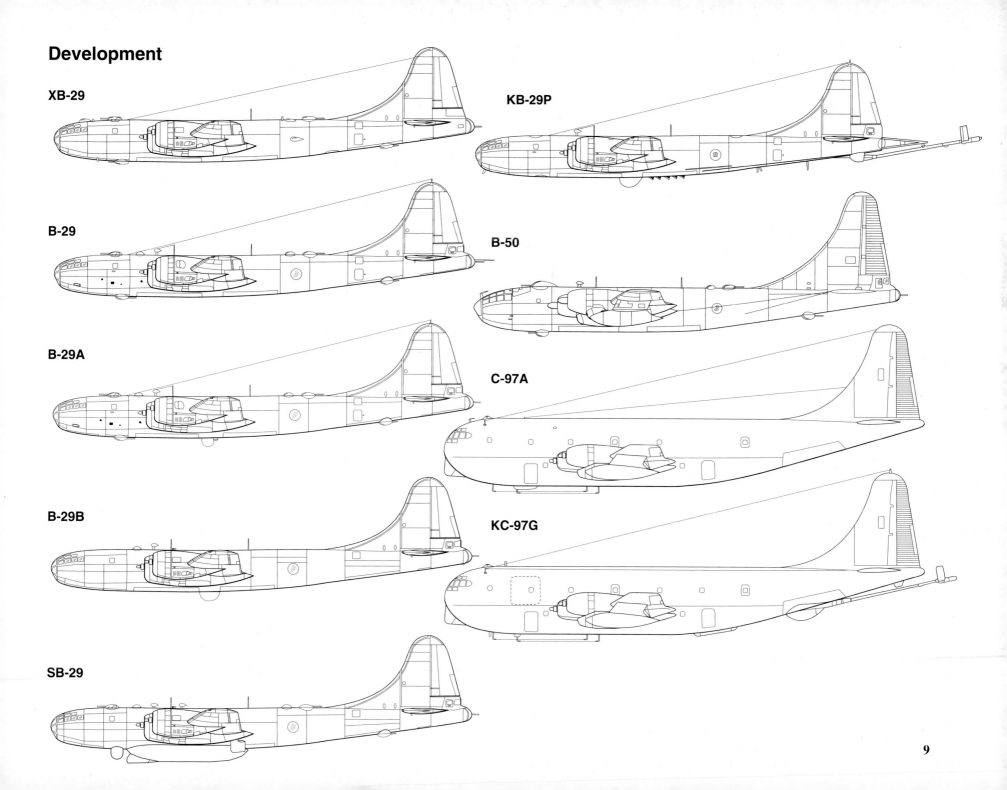

XB-29

B-29

B-29A

B-29B

SB-29

KB-29P

B-50

C-97A

KC-97G

B-29 SUPERFORTRESS

At first sight, the only difference between the XB/YB-29s and the production **B-29** was the change from three bladed propellers to 16 feet 7 inch four bladed Hamilton Standard full feathering propellers. But the additional more subtle changes were exceedingly important. The remote control Sperry turrets with periscopic sighting and teardrop shaped sighting blisters, were replaced with General Electric remote turrets having lead computing gun sights. The teardrop sighting blisters were changed to large circular bubbles in the middle of the aft fuselage. Finally, the engines were replaced with R-3350-13 engines that had reworked superchargers — it had been found that a malfunction in the supercharger was the cause of the Eddie Allen crash. The first aircraft from the Wichita assembly line were delivered in Olive Drab over Neutral Grey paint.

Slowly but surely the bugs were worked out, both on the assembly line and at the modification centers set up to make the changes. Boeing/Wichita built a total of 1620 B-29s, Martin/Omaha built 204, Bell/Marietta built 357, for a total of 2181 B-29s — much too slow to suit the Army brass. As the first production B-29s came off the assembly lines in Wichita on 7 October 1943, and at the Bell plant in Marietta, Georgia, on 30 December 1943, Army was already organizing them into units. The B-29 was such a drastic departure from other bomber types, and so much larger, that the units were designated Very Heavy Bombardment Groups. The first B-29 wing, the 58th Bomb Wing, was initially established at the Bell Marietta factory. The 58th subsequently moved to Smoky Hill Army Air Base in Kansas, with five groups — on paper!

It was here that the 'Battle of Kansas' took place. President Roosevelt had promised China's President Chiang Kai-shek that B-29s would be in place at the new Chinese bases by 15 April

The B-29-1 differed from the XB-29 in having General Electric remote control gun turrets with lead computing gun sights, and four bladed Hamilton Standard constant speed propellers. The first B-29-1 came off the Wichita assembly line during the Fall of 1943. (Peter Bowers)

1944. But the aircraft being delivered to the groups at Smoky Hill were nowhere near being combat ready. The B-29s were slated to leave Smoky Hill on 10 March 1944, but none were ready. 'Hap' Arnold went wild! The B-29s rolling off the assembly lines needed a minimum of fifty four major modifications before they could be considered 'combat ready'. Everything from the fire control system to the engines had problems that had to be fixed, especially the

The B-29-1 had a revised, enlarged tail gunners compartment, shorter rudder, and was fitted with a 20MM cannon in addition to the two .50 caliber guns in the tail. (Peter Bowers)

engines. General Arnold took personal control of the entire B-29 program.

On 26 March 1944, after much butt-kicking and harassing, the first B-29s were enroute to the Far East, via North Africa. They were held up in Cairo for a week due to engine overheating. The outside temperature was 115 degrees, something the engineers hadn't taken into account. And in India, where the airplanes would be based, the temperature often reached 120 degrees! Engine baffles and oil lines had to be re-designed and replaced on the spot. If a change worked, the 'fix' was quickly performed on all other airplanes enroute, and incorporated on the assembly line.

By the 15 April deadline, there were thirty-two B-29s on the ramp at Kharagpur, India, with well over one hundred in place by the end of the month. On 26 April, the first B-29s crossed 'The Hump' (Himalayan Mountains), landing at the specially prepared B-29 landing field at Kwanghan, China. Supplies for this end of the B-29 operations had to be flown in by, what else, the B-29s themselves. Several aircraft were stripped and modified as fuel tankers and cargo haulers to carry the needed supplies over The Hump. On some supply flights over The Hump when heavy head winds were encountered it could take as many as twelve gallons of fuel to deliver one. Supplies of fuel and armament was the linch pin of B-29 operations in China.

Once supplies were in place, the B-29s were ready and anxious to go to war. The first contact with the Japanese actually had come on that first flight over The Hump, when Japanese Army Air Force Ki-43 Oscar fighters jumped one of the inbound B-29s. Hits were scored by both sides, and one Oscar was claimed as shot down. Aboard the B-29, however, an old jinx had reared its ugly head when three of the four turrets jammed. The tail gunner was credited with the Oscar victory.

On 5 June 1944, the first mission was flown. Taking off from Kharagpur, COL Jack Harmon led one hundred B-29s to bomb the Makasan Rail Yards in Bangkok, Thailand. Only eighty aircraft arrived over the target — twenty aircraft were either forced to turn back because of mechanical problems, or crashed enroute. Bombing through overcast, it was later determined that only eighteen bombs had hit the target. A rather inauspicious beginning for the "super bomber". However, that would change rapidly.

On 15 June 1944, the B-29s made their first strike against Japan itself, when they hit the Imperial Iron and Steel Works in Yawata, Japan. It would not be the last. The Twentieth Bomber Command would fly 72 missions in the China/Burma/India theater, one of which was the longest mission ever flown, when thirty-one B-29s left China Bay, Ceylon, to hit the oil refineries at Palembang, Sumatra — a 3900 mile round trip!

Meanwhile, on 15 June 1944, while ninety-four B-29s based in China were attacking targets in Japan, a far more important event was taking place some 1500 miles away — the invasion of Saipan in the Marianas. On 24 August, as the Navy SeaBees were putting the finishing touches to Isley Field, the air echelon of the 73rd Bomb Wing arrived to begin preparations for the arrival of the first 21st Bomber Command B-29s. On 12 October, General Haywood 'Possum' Hansell brought **JOLTIN' JOSIE - THE PACIFIC PIONEER** into Isley Field. 'JOSIE' was the first of the many B-29s to come to the Marianas. Sixteen days later, eighteen B-29s carried out the first attack from the Marianas, hitting the Japanese sub pens in the Truk Atoll. On 24 November, one hundred eleven B-29s left the Marianas and headed north. Their target — the Musashino aircraft factory in Tokyo! It was the first of many B-29 raids carried

Production models of the B-29 were equipped with 16 foot 7 inch Hamilton Standard full-feathering four bladed propellers in place of the 17 foot three bladed propellers used on the XB-29s. (Peter Bowers)

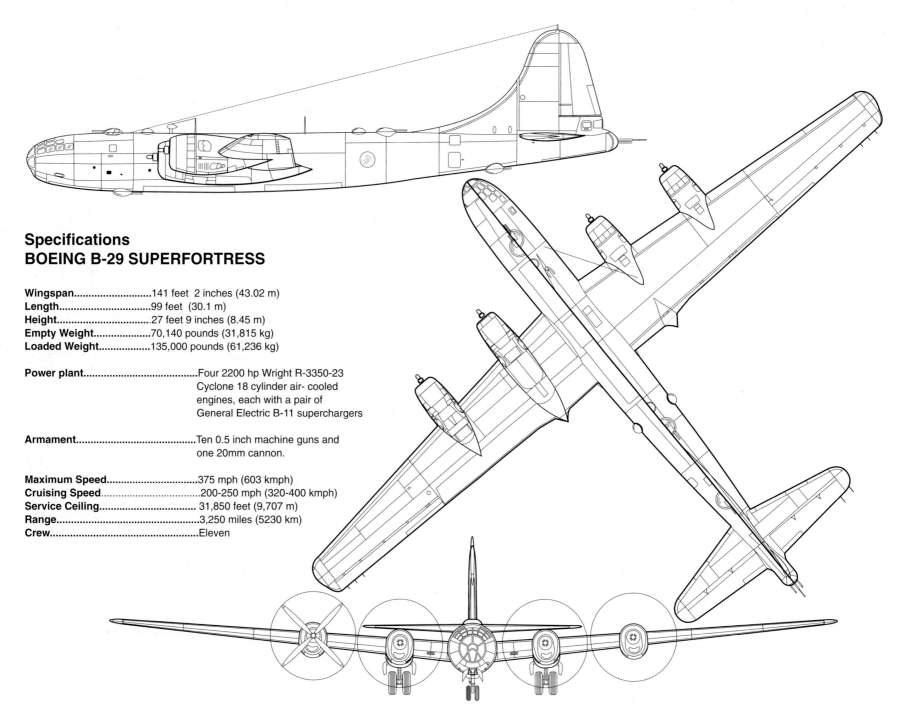

Specifications
BOEING B-29 SUPERFORTRESS

Wingspan............................141 feet 2 inches (43.02 m)
Length................................99 feet (30.1 m)
Height.................................27 feet 9 inches (8.45 m)
Empty Weight....................70,140 pounds (31,815 kg)
Loaded Weight.................135,000 pounds (61,236 kg)

Power plant.........................Four 2200 hp Wright R-3350-23
                                    Cyclone 18 cylinder air- cooled
                                    engines, each with a pair of
                                    General Electric B-11 superchargers

Armament...........................Ten 0.5 inch machine guns and
                                    one 20mm cannon.

Maximum Speed................................375 mph (603 kmph)
Cruising Speed...................................200-250 mph (320-400 kmph)
Service Ceiling.................................. 31,850 feet (9,707 m)
Range..3,250 miles (5230 km)
Crew...Eleven

A line up of B-29-1s at the Wichita factory awaiting installation of their propellers prior to delivery during the late Summer of 1943. All forty of the initial Wichita-built B-29-1s, and some Renton built B-29-1s, were delivered with Olive Drab over Neutral Grey camouflage, and Red-bordered national insignias. (Peter Bowers)

Some of the first B-29s into China were stripped of all but the tail turret, and had large fuel tanks installed in the bomb bay and rear fuselage. It was the only way that fuel and supplies could be hauled 'over The Hump' from Indian bases to the B-29s operating from China. (Dave Lucabaugh)

"DING HOW", a 444th BG B-29 on the ramp at Kwanghan, China on 25 October 1944. On 21 November 1944, DING HOW (42-6225), was forced down at Vladivostok, Russia, becoming one of three B-29s used by Tupolev to create the Tu-4, an exact copy of the Superfortress; a truly remarkable engineering feat. (Dave Lucabaugh)

out on the Japanese capitol and its industry.

On 29 March 1945, unable to adequately supply the B-29s, 20th Bomber Command stood down after 72 missions, closing down B-29 operations in the China/Burma/India theater. The 58th Bomb Wing flew their B-29s from the Indian bases south to the Marianas, where they joined the 21st Bomber Command. It brought the total number of B-29 groups to 20. The stage was now set for the total destruction of the Japanese Empire.

During the ensuing months almost every major city in Japan was hit with high explosives and incendiaries; high explosives to blow things apart and incendiaries to set things on fire. From April 1945 through the end of the war, hardly a day went by when the B-29s weren't in the skies over Japan, either by day or by night, culminating with the 1 August attacks by 836 B-29s. The last of the 190 combat missions from the Marianas was flown on 14 August 1945, when 741 B-29s hit targets all over Japan. The next mission came on 27 August, dropping supplies to the various POW camps. The final statistics for the 21st Bomber Command included 34,000 B-29 sorties flown, dropping 160,000 tons of bombs, with the loss of 371 B-29s.

Remember the 'Almost every major city' phrase? One of those major cities that hadn't been hit was Hiroshima.

Hiroshima would be attacked by specially modified B-29s under Project **SILVERPLATE**. **SILVERPLATE** being the code name for the forthcoming nuclear attacks against Japan. Seventeen Martin/Omaha B-29s, from blocks -35, -40, -45, and -50, were pulled from the production lines at random intervals, but only fifteen were sent to the squadron level.

The SILVERPLATE B-29s were highly modified, which included stripping all but the tail guns from the aircraft, fairing over the sighting blisters, and having fuel injected engines turning Curtis Electric propellers. The bomb bay of each aircraft had several internal modifications made allowing the aircraft to carry a new bomb load, including a new H-frame bomb rack, hoist assembly and sway braces, a British-designed shackle and release unit, and different bomb bay doors. Externally, except for some different antennas, the SILVERPLATE B-29s looked like any another light-weight B-29B, minus the APG-15 gun laying radar on the tail turret.

Throughout 1944 extensive tests at Wendover AAB, dropping elusive 'shapes' were conducted, ending in November. On 17 December 1944, AAF activated the 509th Composite Group, the first unit ever organized, equipped, and trained to drop atomic weapons. Although the group had three squadrons, only one squadron was trained for the atomic mission, the 393rd Bomb Squadron. COL Paul Tibbets Jr. took the 393rd BS from Wendover AAB to North Field, Tinian on 26 April 1945, arriving on 29 May.

Carrying spurious markings of other units on Tinian, the 509th flew practice missions, including many to regular targets during the early summer of 1945. Finally, on 6 August 1945, COL Tibbets flying the **ENOLA GAY**, a Martin built B-29 (44-86292), dropped a 9,000 pound LITTLE BOY atomic bomb on the city of Hiroshima, Japan. Three days later, on 9 August, MAJ Charles Sweeney, flying **BOCKSCAR**, a Martin built B-29 (44-27297), dropped a 10,000 pound FAT MAN atomic bomb on the city of Nagasaki, Japan. Each of these atomic weapons with a yield of some 20,000 tons of TNT, totally destroyed the cities instantly. Japan voted to surrender unconditionally only hours after the Nagasaki attack, but did not officially make it known to its people until 14 August. The two B-29s and their valiant crews had saved millions of lives, both American and Japanese, which would have been lost had an invasion of the Japanese Home Islands been required.

Following the end of World War Two many of the B-29s were sent to vast storage bases to await their fate; a few were cut up for scrap, while others stayed on active duty with the US Air Force. The back bone of the Strategic Air Command's equipment into the 1950s, would be the B-29 until Consolidated B-36 Peacemakers and Boeing B-47 Stratofortresses became available. At that time, the remaining B-29s were sent to the reclamation areas to be cut up for scrap.

But a 'small war' in a place named Korea put a halt to the B-29s being cut up for scrap. For three long years, beginning in late June 1950, B-29s from Kadena, Okinawa and Yokota, Japan, were again involved in a conflict in the skies of Asia, hitting targets in North Korea. See Squadron/Signal publication #6020 "MiG Alley" for further details of the Korean opera-

26 April 1944 -- the first B-29 to have contact with Japanese fighters lands at Hsinching, China after the first Superfortress flight 'over The Hump'. Japanese Oscars scored a few hits, but the B-29 tail gunner claimed the first B-29 kill. (Dave Lucabaugh)

tions. Following the end of the Korean War, SAC rapidly phased out the remaining B-29s in their medium bomber force, in favor of the all-jet Boeing B-47 Stratofortress. The last B-29 of any type, a TB-29 (42-65234) from the 6023rd Radar Evaluation Squadron, was retired from the Air Force on 21 June 1960.

Chinese laborers that helped construct the base at Kiunglai, China, watch as mechanics work on the "PRINCESS EILEEN", a 462nd BG B-29-10. Bases such as Kiunglai were constructed by Chinese laborers using mostly hand tools! (Dave Lucabaugh)

15

A B-29 from the 468th Bomb Group is pulled to its parking spot by a tow tractor at Kharagpur, India. All aircraft from B-29-5 onward were delivered in Natural Metal. The unit marking on the nose of the aircraft is unfinished. (James Eckert)

Ground crew members of the 677th Bomb Squadron work on the nose gear retraction assembly at Kharagpur in 1945. B-29s were based in India and China from March of 1944 through March of 1945, when 20th Bomber Command squadrons relocated to The Marianas. (James Eckert)

Top Turret Development

Early Production Two Gun

Four Gun Turret
B-29/B-29A/F-13

"MARY ANN", a 468th Bomb Group B-29 Superfortress unloads 500 pound bombs on Haito, Formosa on 16 October 1944. The B-29 had a maximum capacity of 10,000 pounds of bombs, the largest bomb load carried by any American aircraft in World War Two. (Ernie McDowell)

16

(Below) A 444th BG B-29 over the cloud-covered Himalayan Mountains (The Hump) on 21 November 1944. The many different 'tones' of natural metal can be seen. (Jeff Ethell)

(Above) B-29s from the 444th BG on the ramp at Kwanghan, China following a mission to Omura, Japan on 25 October 1944. The Black diamond with a Yellow numeral was the Group marking, while the Blue fuselage band and outer engine nacelle stripe are for the 679th Squadron. (Dave Lucabaugh)

KiCKAPOO-II, a 468th BG B-29 tanker with 41 missions over The Hump, is pulled to its parking spot by an Army M3 halftrack. B-29 tankers were devoid of all armament except tail guns. (Dave Lucabaugh)

Looking through the pilots side window at another 468th Bomb Group Superfortress enroute to Anshan, Manchuria on 8 September 1944. The APQ-13 thirty inch radome is extended from the bottom of the fuselage. (Dave Lucabaugh)

B-29A SUPERFORTRESS

The B-29A differed from the B-29 in having a four gun upper forward turret instead of the standard two gun turret seen on the B-29. A second difference which was not visible but equally as important, was the way the wings were attached to the fuselage structure in the B-29A. On previous B-29 models, the wing center section was in two parts, right and left sections that bolted together on the aircraft's centerline, the engine nacelles and outer wing panels were then attached to this bolted together center section. On the B-29A, a shorter one piece wing section was installed through the fuselage and attached. To this was added the outboard wing panels and engines. Besides being easier to manufacture and easier for field maintainance crews to work on or replace, the wing of the B-29B was stronger than earlier wings. A slight reduction in fuel capacity resulted as well as an increase in wing span by a foot. Boeing's new plant at Renton, Washington built at total of 1119 B-29As, all carrying the suffix of 'BN' because Renton was a Navy facility, i.e. B-29A-BN.

A scene that would terrify the Axis leaders — American productivity. Boeing built a total of 3,960 B-29s between September of 1942 and May of 1946, including these Renton built B-29As with the four-gun top turret. (David Menard)

The first B-29A-1-BN being moved from the Renton factory to a barge on the Cedar River, for transport to the airfield across the river. Later a bridge was constructed between the factory and the airfield. (Peter Bowers)

B-29s from the 505th BG taxi to the runway at North Field, Iwo Jima, as personnel from the 15th Fighter Group watch. Iwo became a safe haven for B-29s low on fuel or suffering combat damage and was well worth the cost of taking it. (George Lovering)

The 330th Service Group make adjustments and modifications to the big Wright R-3350 Cyclone engines of new B-29s belonging to the 499th Bomb Group at Isley Field, Saipan in November of 1944. (Dave Lucabaugh)

(Below) This rare photo demonstrates the ability of the B-29 to carry ordnance externally, up to four 2,000 pound bombs, in addition to its normal internal bomb load. (Peter Bowers)

(Below)This 19th BG B-29 carries only the twin .50 caliber machine guns common to 21st Bomber Command aircraft. (Ron Witt)

(Above) THE FLYING 8 BALL, a 9th BG B-29, suffered an inflight emergency and landed at Yontan Airfield, Okinawa following a mission to Tokayama, Japan on 27 July 1945. THE 8-BALL has 13 mission markings and a kill marking. (Dave Lucabaugh)

CITY OF LITTLE ROCK", a 39th BG B-29A with full flaps extended, returns to North Field, Guam following a night mission to Sakai, Japan on 10 July 1945. (Dave Lucabaugh)

21

Turret Development

Four Gun Turret
B-29/B-29A/F-13

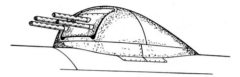

LATE FOUR GUN TURRET
B-29A-40 / B-50 / RB-50

(Above) A J Russell Cheever brought LADY EVE a B-29-5-BA (42-65211) with the 498th BG, back to Isley Field, Saipan with this very large hole in her forward fuselage, after she was hit by some accurate Japanese flak. (Ernie McDowell)

EL PAJARO DE LA GUERRA (THE BIRD OF WAR), a 6th BG B-29 wears the Gloss Black paint adopted for night missions during the Spring of 1945. The Black tail stripe indicates a lead aircraft in the 6th BG. (Ernie McDowell)

499th BG B-29s taxi through the morning mist at Isley Field during the Spring of 1945. Bases in the Marianas were over 1500 miles from Japan, but well within the 3250 mile range of the Superfortress. (Ernie McDowell)

DAUNTLESS DOTTY, an 869th BS B-29 flown by MAJ Robert Morgan of MEMPHIS BELLE fame, was the lead aircraft for the first 21st Bomber Command raid on Tokyo on 24 November 1944. (James V. Crow)

500 pound High Explosive (HE) bombs sit ready for loading aboard a 468th BG B-29 at Pengshan, China. The B-29 could carry a maximum of 10,000 pounds of bombs internally. (Ernie McDowell)

The most classified instrument in the Army Air Force was the famous Norden bomb sight, rarely seen in any photographs. (Larry Sutherland)

B-29B SUPERFORTRESS

The B-29B was built by Bell Aircraft Company in their Marietta, Georgia plant. The B-29B series was in response to MAJ GEN Curtis Lemay's call for a lighter and faster B-29 to carry out low-level attacks on Japanese cities. With Japanese fighter opposition far lighter than anticipated General Lemay wanted his B-29s to go in very low and fast to minimize the threat from anti-aircraft fire — and at night. 21st Bomber Command B-29s would carry large quantities of the new M69 napalm incendiaries to literally burn Japan to the ground. There was only one way to make the B-29s lighter — strip them — which is what General Lemay ordered done to many of the aircraft already in The Marianas.

The Bell-built B-29B was a production stripped variant that met Gen. Lemay's new tactics, but with an improvemet, all upper and lower gun turrets were removed, along with the central fire control system. Only the tail guns remained, but without the 20mm cannon. It was felt that the only fighter threat to the low flying, very fast B-29s would have to come from the rear. The tail mounted twin .50s were aimed and fired with the AN/APG-15B radar fire control system. The AN/APG-15B had a small ball-shaped antenna that hung between the .50 caliber guns in the tail. The weight savings and better aerodynamics of the turretless B-29B took the top speed in excess of 365 mph. Bell built a total of three hundred eleven B-29Bs. The B-29B would be the final assembly line built variant of the B-29, with almost all the B models being assigned to the 315th BW. Many of the B-29Bs were also equipped with the AN/APG-7 EAGLE radar 'wing', which offered the bombardier better definition of the ground terrain during night or during attacks on overcast shrouded Japanese cities.

A 498th BG B-29B on the ramp at Isley Field, Saipan. Almost all of the B-29Bs were assigned to groups in the 315th BW, many of which also had the AN/APQ-7 EAGLE radar wing that provided better target definition. (Ernie McDowell)

Tail Gun Development

Two .50 caliber machine guns and one 20mm cannon   **Two .50 caliber machine guns**   **AN/APG-15B**
TAIL RADOME
two .50 caliber machine guns

DANNY MITE, a 498th BG B-29 (44-69777), unloads 500 pound HE bombs. An intervalometer alternated the bomb drops from both bomb bays to keep the aircraft's center of gravity balanced. (USAAF)

Amiable Amazon was the first B-29 to land at Navy Operating Base, Adak, Alaska, enroute to The Marianas on 10 May 1945. Squadron crews often picked up their aircraft at the factory, applying nose art even before the group markings. (Dave Lucabaugh)

A sight every B-29 crew member welcomed — Iwo Jima based P-51 mustangs appearing in the scanner's bubble as everybody heads for home. Mustangs and Lightnings with drop tanks were the only fighters with long enough legs to fly escort for the Superfortress. (USAAF)

25

The evening of 6 August 1945 — ENOLA GAY sits at North Field, Tinian following the Hiroshima attack. A Martin built B-29-35, ENOLA GAY was devoid of turrets and had a modified bomb bay for carriage of the atomic weapon. (Peter Bowers)

Dave's Dream flew as THE BIG STINK when it was assigned to the 509th BG during World War Two. Dave's Dream dropped the first A-bomb on the Bikini Atoll during Operation CROSSROADS on 1 July 1946. (David Menard)

Carrying large vision domes on top of the fuselage and square windows in the aft fuselage, THE EYE was a specially modified B-29-96-BW used to transport press observers for the Bikini atomic bomb test on 1 July 1946. (David Menard)

TOP OF THE MARK, a 28th BS B-29 at Kadena AB, Okinawa during the late Fall of 1950, carries over 40 mission markers on the nose. The 19th BG, referred to as 'MacArthur's Private Air Force', flew the first B-29 strikes of the Korean War. (Dick Oakley)

Crew chiefs make adjustments to the big Wright Cyclones on MULE TRAIN, a 22nd BG B-29 based at Kadena. The 22nd BG deployed to Kadena in July 1950, when SAC committed four additional B-29 groups to the Korean War. (Joe Bruch)

RESERVED, a 98th BG B-29 based at Yokota AB, Japan in late 1951, wears the Gloss Black camouflage after MiG attacks forced FEAF Bomber Command to fly night missions. (M.H. Havelaar)

FIRE BALL, the first 98th BG B-29 to leave Fairchild AFB for duty in the Korean War. The 98th BG would remain at Yokota, Japan until June of 1952. (Joe Bruch)

(Above) A 92nd BG B-29 at Yokota, being rearmed and refueled for another strike against the North Korean communists. Most strategic targets were destroyed by the early Fall of 1950 with the 92nd returning home in November. (USAF)

(Below) HEAVENLY LADEN flew with the 98th BG out of Yokota. B-29s used during the Korean War were mostly World War Two veterans, dropping 185,000 tons of bombs during the three years of war. (USAF)

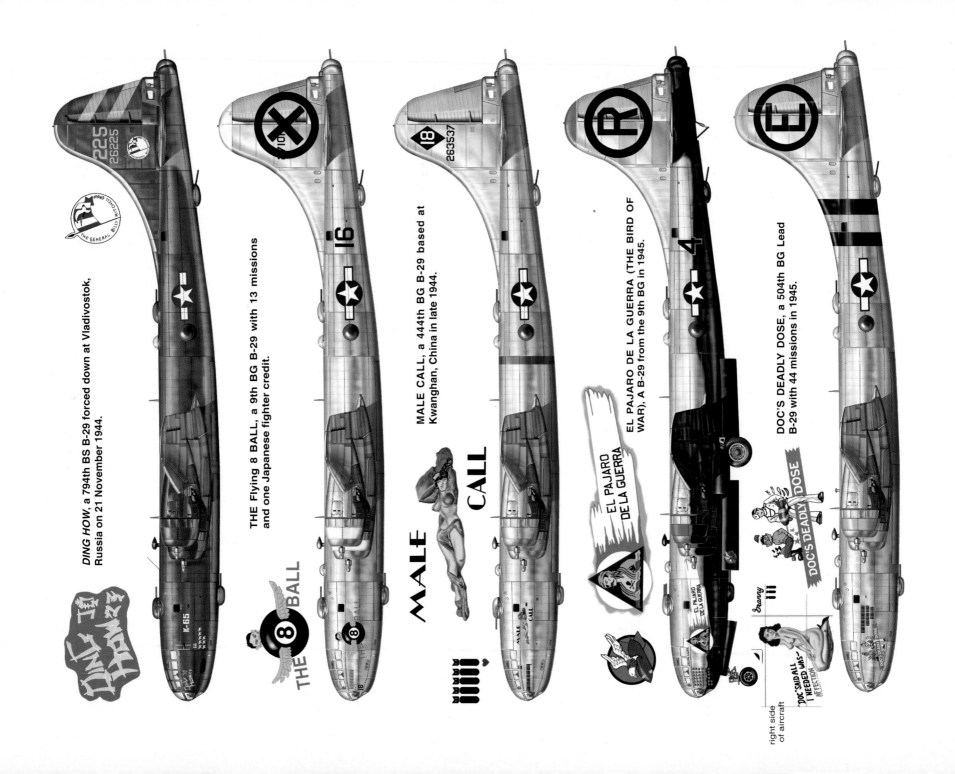

DING HOW, a 794th BS B-29 forced down at Vladivostok, Russia on 21 November 1944.

THE Flying 8 BALL, a 9th BG B-29 with 13 missions and one Japanese fighter credit.

MALE CALL, a 444th BG B-29 based at Kwanghan, China in late 1944.

EL PAJARO DE LA GUERRA (THE BIRD OF WAR), A B-29 from the 9th BG in 1945.

DOC'S DEADLY DOSE, a 504th BG Lead B-29 with 44 missions in 1945.

right side of aircraft

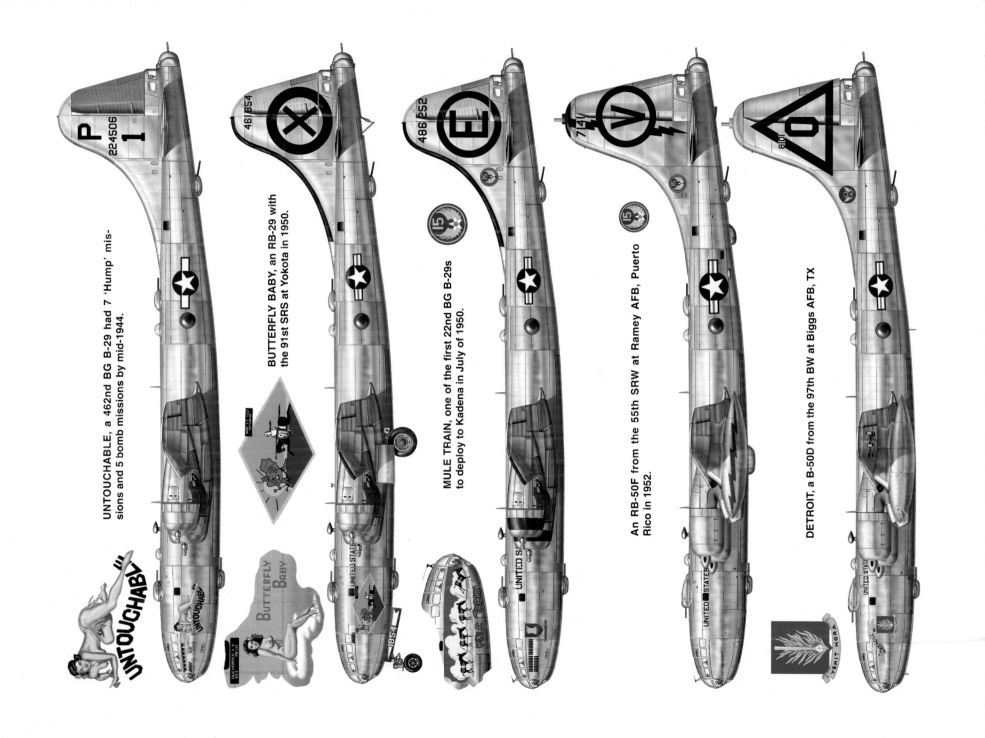

UNTOUCHABLE, a 462nd BG B-29 had 7 'Hump' missions and 5 bomb missions by mid-1944.

BUTTERFLY BABY, an RB-29 with the 91st SRS at Yokota in 1950.

MULE TRAIN, one of the first 22nd BG B-29s to deploy to Kadena in July of 1950.

An RB-50F from the 55th SRW at Ramey AFB, Puerto Rico in 1952.

DETROIT, a B-50D from the 97th BW at Biggs AFB, TX

PACUSAN DREAMBOAT flew 8,198 miles non-stop from Guam to Washington, D.C., DREAMBOAT had the so-called 'Andy Gump' engine nacelles installed which were slated for production on the B-29. The SAC 'Milky Way' has been painted on the nose gear doors. (David Menard)

Equipped with a large radar dish atop the fuselage, this B-29 was a Navy radar airborne early warning picket aircraft and designated P2B-1S. (Joe Bruch)

(Above) This B-29 (45-21793) was assigned to the Army Air Force All Weather Flying Center, flying cosmic ray research missions for the Bartol Research Foundation of the National Geographic Society in April of 1946. The nose and tail markings are Red and Yellow. (Peter Bowers)

A Strategic Air Command (SAC) B-29 of the 307th BG based at Macdill AFB, FL in 1949. B-29s were designated Very Heavy Bombers during WW2, but re-designated as Medium Bombers following the introduction of the much bigger B-36. (Ron Picciani)

A B-29A-75-BN from the 2nd BG being refueled at RAF Lakenheath on 20 August 1948. SAC deployed a number of B-29 groups to bases in Great Britain during the 1948 Berlin Crisis. (Joe Bruch)

Pistol Packin' Mama, a B-29B assigned to the 28th BG of the 15th Air Force at Rapid City, SD, has its buzz number in Red over the Black camouflage. The APQ-15 gun laying radar can be seen silhouetted at the tail turret. (Robert Mikesh)

THE CHALLENGER was the TB-29B assigned to General Jimmy Doolittle when he commanded the 8th Air Force in 1945. In May of 1946 THE CHALLENGER set a record by carrying an 11,000 kg internal payload to 41,561 ft. (Bob Esposito)

The 308th Bomb Wing lines the ramp at Hunter AFB, GA in February 1953. SAC retired the B-29 from service later in the year after the Korean war ended. (Joe Bruch)

Several B-29s were modified for cold weather operations under the designation B-29F, including these aircraft on a snow covered Ladd AFB in Alaska during 1948. (Peter Bowers)

(Below) One of several B-29s that were modified into plush command aircraft and designated VB-29. This VB-29B assigned to SAC has the 'Milky Way' painted on the nose doors and around the rear fuselage. The 0 at the beginning of the serial number means the aircraft is over 10 years in service. (Bob Esposito)

F-13A/RB-29 RECONNAISSANCE SUPERFORTRESS

During the war, it was discovered that there was a need for a very long range reconnaissance aircraft. Both the B-17 and B-24 had been modified for this role, but their range wasn't sufficient for the vast Pacific Theater of Operations (PTO). Army Air Force decided that the B-29 would be the logical choice for this long range reconnaissance mission. Army Air Force pulled one hundred eighteen B-29 and B-29A aircraft from the assembly lines and installed a photo reconnaissance system designed by Air Technical Service Command and Fairchild Photographic Company.

Photo reconnaissance B-29s were designated F-13A (F for photo), no matter what type of B-29 they were based on. All F-13As were rebuilt at the Continental Air Lines Denver Modification Center. Basically, each aircraft had a bank of six cameras installed behind and below the aft crew compartment. There were three K-17Bs, two K-22s, and a single K-18 camera sighting through square windows cut into the bottom and sides of the rear fuselage. Sighting was through a modified B-3 Driftmeter located in the bombardier compartment and operated by the Photo-Navigator.

The F-13As were further modified in having fuel tanks installed in the rear bomb bay. The forward bomb bay could hold either a full load of photo flash bombs, or a cargo platform could be installed which could hold additional film or special cameras. All defensive armament was retained and the standard combat crew of eleven was augmented with the addition of a Photo-Navigator and a Cameraman that maintained and operated the photo systems while inflight.

The first F-13A, TOKYO ROSE, arrived on Saipan on 13 October 1944, flying the first recon mission over Tokyo the same day. The photos proved invaluable during the later attacks against the Japanese capitol. F-13As of both the 1st and 3rd Photo Reconnaissance Squadrons (PRS) operated from both China and The Marianas until the war ended. After the end of the

Camera Installation

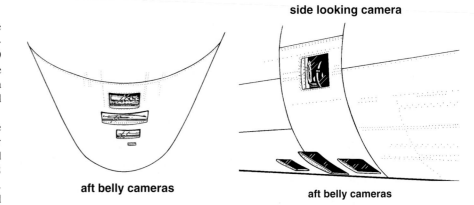

side looking camera

aft belly cameras

aft belly cameras

war many F-13As had their defensive armament removed, with the exception of the tail guns. In 1948, the new US Air Force re-designated all photo reconnaissance aircraft from F for Photo, to R for Reconnaissance and the original aircraft type. Thus the F-13A became an RB-29 or an RB-29A. During the Korean War, the very long range reconnaissance mission was again flown by RB-29s (and some RB-50s) assigned to the 91st Strategic Reconnaissance Squadron (SRS) at Yokota. The final mission of the Korean War was flown by a 91st SRS RB-29A checking North Korean air bases.

Sweet n' Lola was a modified F-13A photo Superfortress assigned to the 509th BG during the Bikini A-bomb tests in July of 1946. Barely seen under the rear fuselage is the indentation of a camera port. (Bob Esposito)

Chat'nooga Choo Choo, a Renton built 468th BG B-29 based in India in 1945, carries 3 photo reconnaissance missions using hand held cameras, ten bombing and nine tanker missions. (James Eckert)

Long range photo reconnaissance missions during the Korean War were flown by RB-29 crews of the 91st Strategic Reconnaissance Squadron (SRS). MOON'S MOONBEAM, an RB-29A (44-61815) is seen at Yokota in 1951. (Tom Mullin)

OVER EXPOSED is perhaps a rather fitting name for a photo reconnaissance aircraft. This F-13A (44-61999) was assigned to the 2nd AF in 1946. (USAF)

FLAK SHACK was hit with flak in March of 1952 while over the Yalu River, but its 91st SRS crew brought it back to Yokota for a crash landing. The RB-29A was stripped of its usable parts and then scrapped. (USAF)

35

58th Bomb Wing, China 1945. (Tom Ivie)

444th Bomb Group, India 1945. (USAAF)

468th Bomb Group, India 1945. (James Eckert)

73rd Bomb Wing, Saipan. (USAAF)

444th Bomb Group, India. (Ernie McDowell)

331st Bomb Group, Guam. (Norris Graser)

771st Bomb Squadron, India. (Ernie McDowell)

497th Bomb Group, Saipan.. (Ernie McDowell)

500th Bomb Group, Saipan. (James Lansdale)

36

497th Bomb Group, Saipan 1945. (USAAF)

462nd Bomb Group, China. (James Lansdale)

462nd Bomb Group, China (James Lansdale)

462nd Bomb Group, China. (James Lansdale)

497th Bomb Group, Saipan. (Norris Graser)

498th Bomb Group Saipan. (Ernie McDowell)

500th Bomb Group, Saipan. (James Lansdale)

500th Bomb Group, Saipan. (USAAF)

499th Bomb Group, Saipan. (USAAF)

37

SB-29 SUPER DUMBO

During the long over-water missions that were being flown from The Marianas, the need for a rescue aircraft that could match the range of the B-29 became obvious. The natural choice for a long range rescue aircraft to match the range of the B-29 was of course, the B-29. Army Air Force had sixteen B-29s modified for the long range rescue mission, by installation of an EDO Corporation 29 foot 9 inch type A-3 lifeboat. These rescue aircraft were re-designated SB-29, and nick-named SUPER DUMBO.

The EDO A-3 lifeboat was carried stern-first, under the fuselage where the bomb bay doors normally would be. The boat was externally mounted to the SB-29, completely covering the bomb bay door area. The bomb bays themselves had extra fuel tanks and/or cargo platforms installed to both extend the range of the SB-29, and to carry additional rescue supplies. Installing the lifeboat over the bomb bay area necessitated moving the APQ-13 radar to the area where the lower forward gun turret had been. All other defensive armament was retained. The SB-29s were assigned to bases that had long over-water flights into or out of that base. During the Korean War, SB-29s from the 2nd and 3rd Air Rescue Squadrons escorted all B-29 missions out of Kadena and Yokota.

EDO A-3 Life Boat

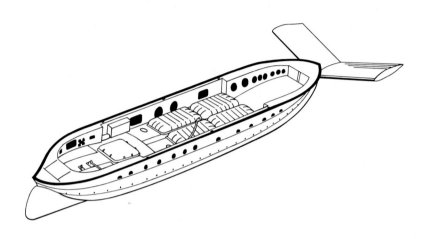

An SB-29 drops an EDO A-3 lifeboat into the Gulf of Mexico during a training exercise in 1952. To make room for the boat the APQ-13 radar was moved to the forward lower turret position. (Joe Bruch)

The EDO A-3 lifeboat hits the water very close to the downed pilot floating in a life raft. The 29 foot 9 inch lifeboat carried survival supplies: food, water, radios, and maps for downed airmen. (Peter Bowers)

38

(Above) An SB-29 of the 3rd Air Rescue Squadron (ARS) over Japan in 1951. SB-29 SUPER DUMBOs escorted all B-29 missions during the Korean War; as such, they carried full defensive armament. (USAF)

(Below) A trio of Air Rescue Service SB-29s at Maxwell AFB, AL in 1951. When a lifeboat is not being carried the bomb bay doors can be opened. Additional fuel and rescue supplies were carried in the bomb bays. (Robert Mikesh)

(Below) An SB-29 from the 2nd ARS taxies to the active runway at Kadena. The A-3 lifeboat and fuselage, wing, and tail stripes are Yellow with Black trim. (USAF)

KB-29 FLYING TANKER

It had long been the dream of air force planners to be able to extend the range of an aircraft through some kind of inflight refueling. The feasibility of inflight refueling had been demonstrated as early as January 1929, when two Army Air Corps Fokkers refueled each other in mid-air during a record breaking 150 hour flight. Towards the end of the Second World War Flight Refueling Ltd, a British company, had developed a workable system of inflight refueling that could be used on different types of aircraft as either tankers or receivers. The Flight Refueling Ltd system employed a hose on a motorized reel with a cone-shaped fueling receptacle called a "drogue" on the end of the hose. The receiving aircraft had a fueling probe and using a grapnel to pull the drogue into the receiver aircraft where the probe could be maneuvered into the cone shaped drogue. Once contact was established and locked, the tanker could begin pumping fuel to the receiver aircraft.

Seeking to extend the range of its B-29 force, Strategic Air Command ordered one hundred sixty-six B-29s modified for inflight refueling; ninety-two as KB-29M tankers and seventy-six as B-29MR receiver aircraft. Boeing Wichita was re-opened to carry out the modifications beginning in 1948. Both tanker and receiver aircraft had similar modifications. All defensive armament (except the tail turret on the B-29MR) was removed. The aft bomb bay of both tanker and receiver was filled with one of the so-called Tokyo Tanks, however, the KB-29M tanker had both bomb bays filled with fuel tanks providing a total of over 1,000 gallons.

The refueling procedure was quite involved and time consuming. The receiver aircraft deployed a line to 'catch' the refueling hose deployed by the tanker, pulling the refueling hose into the receiver aircraft. Once inside, the drogue was joined with the fueling probe, and fuel was pumped from the tanker. All tanks of the receiver aircraft could be topped off, thus extending the range of the B-29MR indefinitely. The KB-29M carried a crew of seven. The first KB-29Ms were delivered to the 43rd and 509th Air Refueling Squadrons (ARefSs) in June 1948.

The Flight Refueling Ltd system was developed for other aircraft types by installing the probe at some point easily seen by the receiver pilot, such as a wing tip fuel tank. Thus the receiver pilot could literally fly the probe into the drogue, achieve contact and begin taking on fuel much more quickly than with the B-29 refueling system. The probe and drogue system was combat tested during the Korean War when several KB-29Ms from Detachment A of the 43rd ARefS deployed to Yokota AB, Japan in November of 1951. As part of Operation HIGH TIDE an RF-80 from the 67th TRW, with probes installed in the nose of its wing tip tanks, was the initial test bed aircraft.

After the successful test with the RF-80 Air Force deployed the 116th Fighter Bomber Wing to the Korean Theater. The 116th FBW F-84Es had refueling probes in the nose of the wing tip tanks. They refueled from the 43rd ARefS KB-29Ms over the Sea of Japan for missions deep into North Korea, missions they would not have been able to fly without inflight refueling. The F-84Es were scheduled to refuel on both the ingress and egress from North Korea. Unfortunately the in-commission rate of KB-29Ms was such that the F-84s usually were forced to spend the night at Taegue.

YKB-29T TANKER

A single KB-29M was further modified with three refueling drogues — one mounted in the normal position under the rear fuselage, and two more mounted inside fuel tank pods under

The upper aircraft is a KB29M tanker of the 509th Air Refueling Squadron that has deployed the fueling hose and made contact with, and is refueling a B-29MR of the 509th Bomb Group. The hose was snagged in mid-air by the B-29MR crew and dragged inside where it was plugged into the refueling receptacle. (Peter Bowers)

HOMOGENIZED ETHYL was a KB-29M assigned to the 43rd ARefS at Davis-Monthan AFB, AZ in 1950. The first KB-29Ms were delivered to SAC in 1948. (David Menard)

the wing tips. Each pod contained a hose reel and fuel drogue assembly. The YKB-29T was used to demonstrate the refueling procedure, which could be used with other NATO types such as the Gloster Meteor. This system was further developed and operationally used on the Tactical Air Command's KB-50 tanker fleet.

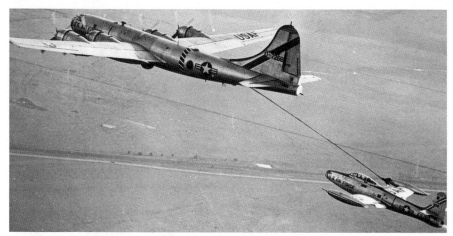

A KB-29M belonging to the 421st Air Refueling Squadron (ARefS) makes contact with an F-84E of the 49th Fighter Bomber Group (FBG). The F-84E has a refueling probe in the nose of its wing tip tank, which the pilot flew into the drogue hose trailing behind the Superfortress tanker. (Merle Olmsted)

KB-29P & THE FLYING BOOM

Although effective, the probe & drogue system suffered from a number of limiting factors. Making contact in heavy weather, or at night, was extremely difficult and the drogue hose was subject to whipping in turbulent air. Boeing sought to develop a better system using a rigid metal pipe to transfer the fuel.

The transfer pipe commonly known as the Boeing Flying Boom, was telescoping in length, and had small winglets near the end to provide lift. The end of the boom had the fueling probe. The receiver aircraft had a refueling receptacle called a slipway, which was built-in and covered by doors. The slipway could be mounted anywhere on the receiver aircraft. The boom operator in the tanker flew the boom into the open slipway of the receiver aircraft. On B-47s and F-105s the slipway was in front of the cockpit, on F-4s and B-52s the slipway was behind the cockpit, and F-84Gs mounted the slipway in the upper port wing.

The boom operator in the tanker sat in the former tail gunners position, looking out through a sighting blister mounted in place of the tail turret. First he guided the receiver aircraft into position by using both voice signals and a lighting system on the underside of the rear bomb bay doors. Once in position, the boomer flew the transfer boom into the open slipway of the receiver aircraft, usually accompanied by some very suggestive noises emanating from both aircraft.

The Boeing plant at Renton, Washington, which had re-opened for production of the C-97, was chosen to modify 116 B-29s with the Boeing Flying Boom under the designation KB-29P. All armament and self-sealing fuel tanks were removed, being replaced by standard non-self-sealing types, which held a few more gallons of fuel. Bomb bay fuel tanks were installed, bringing the total to 12,000 gallons. The Boeing Flying Boom assembly was installed under the extreme aft end of the fuselage, and was attached to a metal quadrapod, which held the boom in a rigid position during non-refueling air and ground operations. The KB-29P carried a crew of nine. The first KB-29Ps were delivered to Strategic Air Command on 1 September 1950.

SAC KB-29Ps were involved in two major trans-Pacific crossings, as well as combat refueling operations during the Korean War. Operations FOX PETER ONE and FOX PETER TWO, sent the 31st and 27th Fighter Escort Wings to bases in the Far East in 1952. The F-84Gs made the 11,000 mile flight by inflight refuelings from KB-29P tankers that were waiting along the route. Detachment A of the 43rd ARefS had several KB-29Ps to refuel the RB-45C Tornados, flying high speed reconnaissance missions near the Yalu River (and beyond). A total of eleven SAC squadrons were equipped with KB-29Ps before the type was phased out in favor of KC-97 and KC-135 jet tankers in 1957.

Members of Det 4 of the 43rd ARefS meet with pilots from the 116th Fighter Bomber Group at Taegu AB in February of 1952 to discuss Operation HIGH TIDE, the first combat aerial refueling of a strike force in history. The KB-29Ms refueled the bomb laden F-84E Thunderjets over the Sea of Japan. (Joe Bruch)

The KB-29P used the Boeing developed Flying Boom system, a telescoping metal fuel transfer pipe with small winglets that the boom operator flew into the receptacle on the receiving aircraft. An additional radome is carried on the rear spine of these 509th ARefS KB-29Ps. (Boeing)

41

(Above) KB-29Ps from the 91st ARefS enroute to a refueling track over the Yellow Sea in 1954. SAC KB-29Ps refueled B-47s, B/RB-45s, and F-84Gs during the mid-1950 (Boeing)

(Below) Operation FOX PETER ONE sent the 31st Strategic Fighter Wing to the Far East, refueled by 91st ARefS KB-29Ps at various points along the route. (Boeing)

42

(Above) A SAC KB-29P from the 509th ARefS sits on a snowy ramp at RAF Lakenheath in 1954. The KB-29P held over 12,000 gallons of JP-4 jet fuel, in addition to its normal internal tankage of aviation gas to feed its own big Wright Cyclone engines. (Joe Bruch)

(Below) An RB-45C from the 91st Strategic Reconnaissance Squadron hooks up to a 91st ARefS KB-29P over Japan during the Korean War. The RB-45C needed inflight refueling to reach its assigned target areas on or above the Yalu River. (USAF)

Refueling

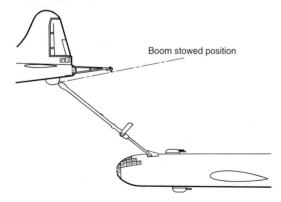

Boom stowed position

EB-29 MOTHER SHIP

The power and robustness of the B-29's engines and airframe was such that it was used to conduct a number of experimental projects no other aircraft in the Air Force's inventory could perform. In a joint venture between the National Advisory Committee for Aeronautics (NACA) and Bell Aircraft Corp, a superfortress,under the designation EB-29, was modified to carry aloft the Bell XS-1 in which Chuck Yeager completed his record breaking flight on 14 October 1947.

The so-called EB-29 Mother Ship was used to carry aloft the XF-85 Goblin parasite fighter on a trapeze mechanism mounted in the bomb bay which could both launch and recover the tiny experimental fighter. Successful launching trials were conducted by an EB-29A (44-84111) in July of 1948, with the first successful recovery in October. If the program had gone further than these experiments it would have been the B-36 Peacemaker, the successor of the Superfortress, to mount the Goblin.

(Right) The X-1A and X-1B were carried aloft by the same EB-29 that carried Yeager and the XS-1. The bomb bay area has been cut away to carry the rocket plane in a semi-internal position. (Bob Esposito)

(Below Right) An EB-29 (E for Exempt) carries the Bell XS-1 research aircraft aloft during one of the many test drops conducted before MAJ Chuck Yeager's record-breaking flight. The elevator is in place over the XS-1s pilot entry door. (Peter Bowers)

(Below) The Navy had a similar rocket research program with the Douglas D-558-2 Skyrocket, which was carried aloft by a modified B-29, under the Navy designation P2B-2S. The D-558-2 was the first aircraft to crack the double sonic barrier: Scott Crossfield was the pilot. (Nick Williams)

TB-29 TRAINER/TARGET TUG

The TB-29 designation was used for two different B-29 missions — as a flying trainer aircraft and as a target tug. In the trainer role, the TB-29 was used as in-flight trainers for all the various crew members, from the aircraft commander to the gunner. Air Defense Command also used several TB-29s to develop and test interceptor tactics used by its fighter squadrons. These were known as Radar Calibration Flights. TB-29 trainers were equipped with additional seats for use by instructor and evaluation crew.

The second TB-29 designation was used to denote B-29s that were modified with tow target systems. These aircraft had electric reels installed in the aft bomb bay that could deploy a cable well over a mile long, minimizing the threat of the tug being hit during target practice. Attached to the end of the cable was a target sleeve or "rag". These target sleeves were mounted under the empty tail gunners position in a long box-like pod. Two rags were carried in each target pod, each being connected to a separate reel in the aft bomb bay. The tow target operator sat in the right scanners station. TB-29 target tugs were used at any live-fire exercise, such as the Air Defense Command World Wide Rocket Competition held at Yuma, AZ in the mid-1950s.

(Above) A Randolph Field TB-29 flies formation with a pair of AT-6 trainers during a simulated gunnery exercise in the late 1940s. (David Menard)

(Left) A TB-29 from Randolph AFB over Texas during the late 1940s. TB-29s were crew trainers for all aspects of the B-29 mission, including pilots, gunners, and navigators. (David Menard)

(Below) The second TB-29 designation in use indicated a target tug aircraft, which carried a pair of tow target sleeves in the box mounted under the rear fuselage. These Black camouflaged TB-29s "pulled the rag" during the 1953 FEAF Gunnery Competition. (Ron Picciani)

WB-29 WEATHER RECONNAISSANCE

WB-29s were standard bomber aircraft modified for the weather reconnaissance mission. Initially stripped of all defensive armament, the WB-29s had meteorological recording devices like the Psychonometer that measured water vapor content, an aerograph to measure relative humidity, several radio altimeters, and a Radiosonde unit. The Radiosonde picked up signals from parachute-dropped transmitters that recorded temperature, humidity, and air pressure data. Commonly referred to as "Hurricane Hunters", the WB-29s would routinely penetrate the 'eye' of a hurricane to record data and attempt to predict the path of the storm. The first B-29 penetrated a hurricane on 7 October 1946.

But hurricane hunting wasn't the only way that WB-29 crews were put in harms way. During the Korean War, WB-29s from the 56th and 512th Strategic Reconnaissance Squadron/(Weather)Very Long Range, flew every day above the 38th Parallel to report weather conditions over North Korea. The Atomic Energy Commission also routinely used Air Weather Service WB-29s to monitor airborne radiation levels resulting from the surface nuclear tests conducted from 1946 on. First introduced to Air Weather Service in 1946, the WB-29s were flown into the late 1950s before being replaced by WB-50 aircraft.

(Right) SURPRISE PACKAGE, a WB-29 from the 512th Reconnaissance Squadron,Weather(VLR), during Operation BUZZARD, flew weather reconnaissance missions over North Korea beginning on 26 June 1950. (USAF)

(Below right) A WB-29 weather crew of the 53rd SRS getting a last minute briefing at Kindley AFB, Bermuda. Their mission will be to penetrate the eye of the hurricane, and document wind velocity, air pressures, and other weather data. (USAF)

(Below) A WB-29 from the 388th WRS, Air Weather Service on the ramp at Kadena AFB in 1952. Known as Hurricane Hunters, the WB-29s tracked both hurricanes and typhoons throughout the world. This WB-29 retains its tail guns in the hostile area of the Far East. (Bob Esposito)

THE TUPOLEV Tu-4 SOVIET SUPERFORTRESS

On 29 July 1944, some seventy B-29s of the 58th Bomber Wing, took off from Chinese bases to hit the steel plant in Anshan, Manchuria. One of them ran into some accurate flak which punctured the fuel tanks. Running desperately short of fuel the big bomber was forced to land at Vladivostok, Russia. The crew was returned, but the airplane was not. Two more B-29s were forced down at Vladivostok in November, with the same results. The three interned B-29s were flown to Moscow where Andrei Tupolev was given the less than enviable task of copying the huge aircraft, right down to the last rivet.

At the 3 August 1947 Soviet Aviation Day display at Tushino Air Base near Moscow, the U.S. found out what happened to its aircraft when three B-29s and a transport aircraft that looked very much like a B-29, made a fly-by debut. Under the designation Tu-4 Tupolev had succeeded in copying the Boeing B-29 Superfortress and had developed the Tu-70 transport using the B-29 wing and tail planform. The Tu-4, NATO code-name BULL, even had copies of the Wright R-3360 engines —Shvetsov ASh-73TKs.

Theft? Certainly! But also an engineering feat of enormous proportions that would provide the Soviet Aviation industry with a leap forward in the technological race in producing long range aircraft.

The first Tu-4 had been completed and test flown during the summer of 1946. Tupolev eventually would build over 1200 Tu-4s that were identical to the Boeing B-29. The basic design of the B-29 was modified in 1946 to create a large transport aircraft under the designation Tu-70. Tupolev had combined a longer and wider fuselage, a typically Tupolev stepped nose, with the wings and tail of the Tu-4, and powered by the same ASh-73TK engines. Capable of carrying 70 passengers, the fuselage was 17 feet longer and 2 feet wider than the Tu-4 bomber. To be built as a commercial transport for Aeroflot, the Tu-70 was suddenly canceled without going into production.

However, Tupolev developed the type into a military transport, with large doors under the aft fuselage. Under the designation Tu-75 it had upper and lower twin gun turrets, plus a two-gun tail turret. The Tu-75 was further developed into the Tu-80 and Tu-85 bombers, with increased horsepower, longer wings and fuselage. Work on the Tu-85 was discontinued in favor of the turboprop powered Tu-95, the prototype of the excellent Tu-20 BEAR.

The Chinese were supplied with some 100 Tu-4 bombers, at least two of which were re-powered with Ivchenko turboprop engines. One of these is on display at The Museum of Chinese Aviation in Beijing.

DING HOW (42-6358) at the Tupolev factory in 1945 after having been forced down at Vladivostok on 21 November 1944. The American aircrew was eventually returned, but the aircraft was not. (Hans-Heiri Stapfer)

(Above) The RAMP TRAMP, a 771st BS B-29, was forced by low fuel to land at Vladivostok, Russia in November of 1944. It was one of three used by Tupolev to build an exact copy for use by the Soviet Air Force. (Hans-Heiri Stapfer)

(Below) A lineup of Tu-4 bombers, code named BULL by NATO, is seen at an airfield outside of Moscow in the mid-1950s. Tupolev built over 1200 Tu-4s, the first long range bomber aircraft in the Soviet Air Force. (Hans-Heiri Stapfer)

A Soviet aircraft commander addresses the crew of his Tu-4 prior to a mission. Minor differences include a different type of antenna as well as the location of the horseshoe antenna on the under side of the fuselage. (Hans-Heiri Stapfer)

The Tu-70 was a commercial transport aircraft based on the Tu-4 bomber; it had B-29 wings, engines, and tail assembly, the completely new fuselage had a stepped nose. The glass nose cone gave it the more ominous look of a bomber rather than a transport. (Hans-Heiri Stapfer)

The Tu-80, the successor to the Tu-4 strategic bomber, with a stepped nose and wings moved to the middle of the fuselage is beginning to take on a definite Tupolev look. (Hans-Heiri Stapfer)

The Tu-85 was the ultimate in the basic Tu-4 development, with a longer fuselage and wingspan, and powered by 4300 hp VD-4K engines. Top speed was 404 mph, with a range of 12,200 km. On very short missions, a 44,000 pound payload could be carried. (Hans-Heiri Stapfer)

WASHINGTON B MK I

During the uneasy peace of the immediate post-war years the RAF had only one aircraft capable of long range strategic strikes against Soviet targets, the AVRO Lincoln, a development of the veteran Lancaster heavy bomber. However, the Lincoln was terribly outmoded in the day of the MiG. So was the B-29 for that matter, which was painfully shown in the skies over Korea. But the B-29 was the only nuclear attack aircraft available for use by the RAF.

On 27 January 1950, the US Defense Department loaned the RAF three B-29 and eighty five B-29A bombers, which the RAF designated WASHINGTON B Mk. I. On 20 March 1950 British Ambassador Sir Oliver Franks formally accepted the aircraft transfer. Crews from the USAF 307th BG delivered the first four aircraft to RAF Marham on 22 March. The RAF Washingtons were in service for three years, equipping a total of ten squadrons counting the conversion unit. By 1954 the remaining flyable Washington's (70) were returned to the US, having been replaced by Canberra jet bombers.

Two Washingtons were in Royal Australian service in 1952 where they remained until 1957 when they were sold for scrap.

(Right) The WASHINGTON B Mk Is were USAF B-29s loaned to the RAF during the post war years until the UK could get its long range bomber program underway. (Peter Bowers)

(Below right) A WASHINGTON B Mk I sits in the Arizona desert amid USAF B-29s following its return from service with No. 44 Squadron in March of 1954. (R. Johnson)

(Below) This WASHINGTON B Mk I (WF 512), an ex-B-29A (44-62016) served with No. 44 (Rhodesia) Squadron in 1951, based at RAF Marham. Nine RAF squadrons were equipped with WASHINGTONs. (Jeff Ethell)

B-50 SUPERFORTRESS

As a result of Army Air Force requirements to update and improve the performance of the B-29, a new powerplant — the Pratt & Whitney R-4360 Wasp Major, a 28 cylinder, twin row radial engine rated at 3300 hp was selected to replace the Wright R-3350 currently powering the B-29 which developed only 2200 hp. A second major improvement was the use of the new 75ST aluminum structure in the wing, which made the wing 16% stronger and 600 pounds lighter, and a re-designed tail assembly. Work was begun in 1944 with the improved Superfortress being designated B-29D. Unfortunately the end of the war brought a reduction in all military orders, including those for the B-29. Orders for the B-29D were slashed from 200 to 60 aircraft, with the distinct possibility that all would shortly be cancelled. Under the designation XB-44 a single B-29A was allocated to Pratt & Whitney as a flying test bed for the Pratt & Whitney R-4360 engine. The XB-44 made its first flight in May of 1945.

B-50A

With Congress cancelling all wartime orders for existing aircraft types, the only way to insure production of the improved Superfortress was to make it appear to be a new aircraft type. In December of 1945 Army Air Force redesignated the improved B-29D to B-50, a new aircraft not involved in WW2 production. It was not a complete lie, the B-50 had a new wing structure, completely new engines, and a new tail assembly — more than half of the original design was to be replaced with new components.

Besides the change in wing structure the B-50 also had much larger flaps in order to shorten the take-off and landing distance. The vertical fin and rudder was five feet taller than the B-29. The fin was hinged at the fuselage juncture to permit it to fold down horizontally and fit into existing hangers. Other changes included hydraulic rudder boost and nose wheel steering, improved main gear retraction gear and de-icing equipment. The R-4360 engines had reversible pitch propellers which increased braking capabilities. Other features found on late block B-29s such as the four gun aerodynamic upper turret, and the APQ-13 radar were included on the B-50.

The first production Superfortress had been designated B-29, however there was no B-50. The first of the new bombers rolled off of the assembly line with the B-50A designation. The first flight took place on 25 June 1947. The initial production order called for sixty B-50As, but the last aircraft was built as the YB-50C prototype for the B-54, which was cancelled in April 1949. Deliveries of the B-50A to SAC began in late 1947.

Performance of the B-50A was increased over that of the B-29, but not as great as a disappointed Air Force had hoped, considering the much greater power and lighter wing structure. The problem was an overall increase in weight due to the taller tail assembly and the larger flaps. Empty weight had gone from the B-29's 69,610 pounds to the B-50A's 81,050 pounds and gross weight jumped from the B-29's 105,000 pounds to the B-50's 168,708 pounds. Top speed did increase 20mph to 385 mph, and the service ceiling increased over 5,000 feet to 37,000 feet. But unrefueled range actually dropped by 1,200 miles, from the 5,800 miles of the B-29 to the 4,600 miles of the B-50A.

The range of the B-50A was increased by the use of inflight refueling from KB-29M tankers. All fifty seven B-50As were modified at Boeing Wichita to use the Flight Refueling Ltd. system, with the receptacle installed under the right stabilizer. SAC demonstrated it could live up to its boast of hitting a target anywhere in the world, by flying around the world —

Engine Development

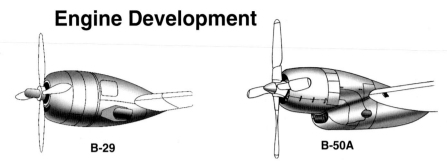

B-29 **B-50A**

nonstop! LUCKY LADY II, a B-50A (46-10) took off on 26 February 1949 from Davis-Monthan AFB and returned on 2 March after 23,452 miles and slightly more than 94 hours. It took four inflight refuelings from eight KB-29Ms to complete the mission.

B-50B

The second B-50 contract called for forty-five improved B-50Bs. The improvements were all internal. However, all but two of the B-50Bs were subsequently sent to Wichita for conversion to RB-50 reconnaissance aircraft with a photo reconnaissance package being installed in the rear bomb bay. Other modifications included installation of an inflight refueling system, and the addition of a pair of 700 gallon fuel tanks mounted under the outer wings.

The forty-three aircraft were converted to RB-50 standards in three variations — fourteen aircraft had special photographic equipment in four camera stations under the re-designation RB-50E; fourteen aircraft had SHORAN radar navigational equipment added under the re-designation RB-50F. Finally, fifteen aircraft had additional radar added and the new one-piece molded bombardier's perspex installed under the re-designation RB-50G.

Following the phaseout of the piston engined RB-50 Superfortress in the reconnaissance role in favor of the jet powered RB-47 Stratojet, the RB-50s were converted to KB-50J tankers. The KB-50J had the triple probe and drogue which had been developed in the YKB-29T, and could refuel three aircraft at the same time. In addition, the KB-50Js had General Electric J47-GE-23 auxiliary turbojet engines mounted on wing pylons. The addition of the jet engines brought top speed up to 444 mph, making it much better suited for refueling TAC

The first B-50A is pulled to a parking spot during roll out ceremonies at Boeing's Seattle factory in June of 1947. The B-50A had the new Pratt & Whitney 3500 hp R-4360-35 engines and a vertical tail assembly five feet taller than the B-29. (Peter Bowers)

fighter aircraft such as the F-100 and F-105. One hundred twelve B-50A and RB-50 aircraft were converted to KB-50J standards.

B-50D

The B-50D was the final bomber variant produced from the original Boeing Model 345 (B-29) design. Two external significant changes were made to the B-50D. First was the installation of a one-piece Plexiglas perspex cone over the bombardier nose replacing the seven piece nose cones on all previous B-29 and B-50 variants. The second change was the installation of an inflight refueling slipway receptacle in the upper forward fuselage, between the canopy and the upper turret. Refueling was accomplished with the Boeing Flying Boom system. The upper turret had to be rotated 90 degrees for the boom to enter the slipway. Two hundred twenty-two B-50Ds were built with the last being retired in 1955.

The final production variant was the TB-50H, a radar bombing trainer for B-47 Stratojet crews that would be using the new K-system. The TB-50H had an additional crew and an electronics station in the aft bomb bay. These were bomb-scoring trainers that emitted a tone at the theoretical release point, which was then plotted by instruments on the ground. Boeing built twenty four TB-50Hs. All of the TB-50Hs were converted to KB-50K tankers, with the same equipment used on the KB-50J.

Two B-50Ds were modified as prototype aircraft for the KB-50 under the designation KB-50D. Eleven B-50Ds were converted to TB-50D trainers. Finally, thirty six B-50Ds were stripped of all offensive and defensive weapons systems, and had meteorological equipment -- Doppler radars, and air samplers installed. These aircraft flew the 'Hurricane Hunter' mission under the designation WB-50D. They were retired in 1967.

The B-50 Superfortress was involved in three wars — Korea, Vietnam, and the Cold War. Several RB-50s were attached to the 91st Strategic Reconnaissance Squadron at Yokota during the Korean War flying long range recon missions to China and the Soviet Union. At least one RB-50 was shot down by Soviet MiG fighters in 1955 and at least six B-29s or RB-50s were shot up by MiGs, flying recon and elint missions along the Iron Curtain in Eastern Europe. During the early stages of the Vietnam War, PACAF fighter aircraft were refueled by TAC KB-50Js both inbound and outbound from targets in North Vietnam including some over hostile territory. The KB-50 tanker fleet was retired in late 1965 in favor of KC-135 jet tankers,

A B-50A over Mt. Rainier in Washington during 1948. Installation of the more powerful Pratt & Whitney engines brought about a very distinctive redesign of the engine cowlings with their much enlarged oil cooler/intercooler intake. The BK buzz number can be seen under the wing. (Peter Bowers)

LUCKY LADY II (46-10), a B-50A of the 43rd Bomb Wing, using inflight refueling flew non-stop around the world between 26 February and 2 March 1949, flying 23,452 miles in 94 hours. The distinctive corn-cob' cylinder configuration of the P&W R-4360 engines is obvious when the cowling is removed for maintainence. (USAF)

This B-50A (46-37) of the 28th Bomb Wing on the ramp at Bolling AFB looks to be the feature of an Air Force open house during May of 1948. Boeing built a total of fifty-nine B-50As. (Ron Picciani)

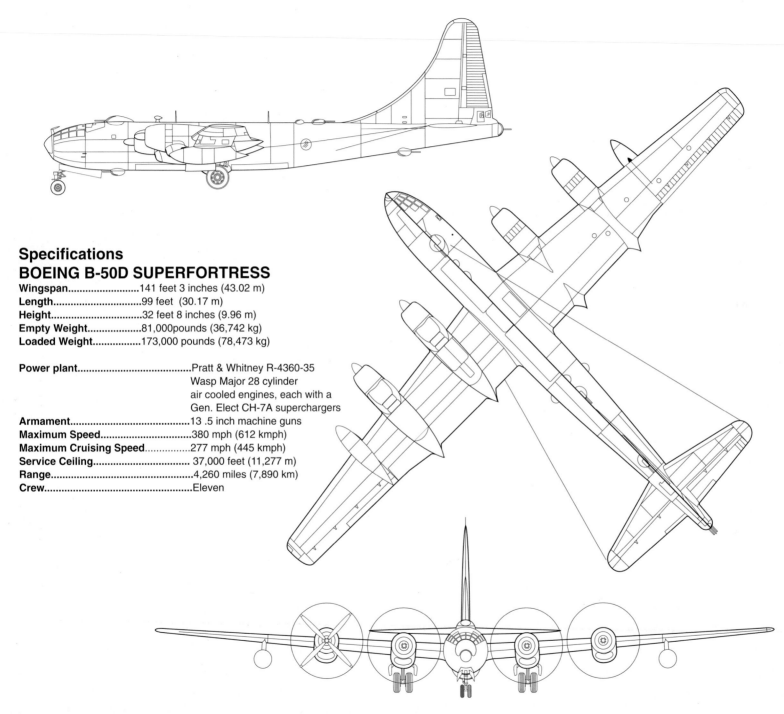

Specifications
BOEING B-50D SUPERFORTRESS

Wingspan..........................141 feet 3 inches (43.02 m)
Length..............................99 feet (30.17 m)
Height...............................32 feet 8 inches (9.96 m)
Empty Weight...................81,000pounds (36,742 kg)
Loaded Weight.................173,000 pounds (78,473 kg)

Power plant..Pratt & Whitney R-4360-35
        Wasp Major 28 cylinder
        air cooled engines, each with a
        Gen. Elect CH-7A superchargers
Armament..13 .5 inch machine guns
Maximum Speed...............................380 mph (612 kmph)
Maximum Cruising Speed................277 mph (445 kmph)
Service Ceiling................................. 37,000 feet (11,277 m)
Range..4,260 miles (7,890 km)
Crew..Eleven

An RB-50E (47-131) resplendent in its Gloss Black lower surfaces and a false serial number on the tail is assigned to the 91st Strategic Reconnaissance Squadron at Yokota AB, Japan in 1953. RB-50s were very active in the Korean War combat zone, with at least one being shot down by MiGs over the Sea of Japan. (Bob Mikesh)

An RB-50F from the 55th Strategic Reconnaissance Wing based at Forbes AFB, Kansas in 1953. The RB-50F had additional ELINT antennas, including a large radome under the rear fuselage. (Andy Meyers)

(Below) The streamlined four gun top turret was adopted for the late production B-29A and all variants of the B-50. The open inflight refueling door can almost be seen in front of the turret. (Peter Bowers)

53

MAC'S EFFORT, a 55th SRW RB-50F flown by LT James McKeown during 1954, carries the Red wing tips and tail markings for operations over snow covered terrain. Three additional antennas can be seen on the spine of the fuselage. (Andy Meyers)

Specially built work stands were developed for the B-29 and were put to good use when B-50 engine work had to be performed in the open. This 97th Bomb Wing B-50D may be a very early production D series since it has the multi-piece bombardier nose. (Joe Bruch)

Detroit, a B-50D from the 97th Bomb Wing at Biggs AFB, TX. The B-50D featured the one piece bombardier nose perspex, and an In Flight Refueling receptacle behind the canopy to connect with the Boeing Flying Boom. A shark mouth has been painted on the wing tank. (Peter Bowers)

Inflight refueling of the B-50D required that the top turret be rotated 90 degree so the guns cleared the slipway for connection of the Flying Boom. (Peter Bowers)

(Above) The KB-50J tanker had three inflight refueling systems — one in the rear fuselage, and one in each wing tip tank assembly. KB-50Js were delivered to Tactical Air Command beginning in 1958. (Peter Bowers)

(Below) *Miss* BEA was a KB-50J assigned to the 421st Air Refueling Squadron in 1964. Many KB-50Js had General Electric J47 jet engines installed on the out board wings in order to increase their speed, making them more compatible with the TAC jet fighters they had to refuel. (E.M. Sommerich)

A Tactical Air Command KB-50J refuels three 322nd Tactical Fighter Wing F-100Cs during Operation MOBILE BAKER, that sent several USAF fighter wings to bases in North Africa. (Peter Bowers)

A pair of 335th TFS F-105Ds being refueled by a KB-50J during Operation ROLLING THUNDER in 1965. Several KB-50Js came under fire while refueling thirsty Thuds coming home after strikes in North Vietnam. (Joe Michaels)

A WB-50 on the ramp at Ramey AFB, Puerto Rico in March of 1964. The WB-50s were all converted from bomber or photo reconnaissance aircraft. (Andy Vies)

A WB-50 from the Air Force Air Weather Service on the ramp at McGuire AFB. Equipped with meteorological instruments, the WB-50 Hurricane Hunters routinely penetrated the 'eye' of violent storms. (Peter Bowers)

C-97 Super Transport

With design work begun as early as 1942 the C-97 was a super transport based on the B-29 Superfortress. The B-29 wing, tail assembly, and the Wright R-3350 engines were retained for the transport, however, a double lobed fuselage with a double decked interior was developed with the lower lobe being the same diameter as the B-29 fuselage with the upper add-on lobe being much larger. Large clamshell doors were installed in the rear fuselage, with hydraulic ramps for ease of access. The C-97 could easily hold 134 troops, three 2 1/2 ton trucks, or a pair of light tanks.

First flown on 9 November 1944, performance of the XC-97 was on par with its B-29 bomber cousins — top speed of 383 mph, service ceiling of 29,000 feet, and a range of 3100 miles. Three XC-97 prototypes were built, followed by six YC-97 service evaluation transports with the so called Andy Gump engine nacelles.

Based on the B-50 the YC-97A, had the wing and tail assembly of the B-50, Pratt & Whitney R-4360 engines and the 75ST aluminum structure wing. Three YC-97As were built, with all being brought to C-97A standards later. The C-97A production machine, differed only in having the AN/APS-42 search radar in a small chin radome.

It was the aerial tanker variant that really made the C-97 famous. Equipped with an improved Boeing Flying Boom, the KC-97 tanker quickly became the the backbone of the SAC tanker fleet during the 1950s. The KC-97E equipped with the new flying boom had four additional fuel bladders mounted on the cargo floor. Boeing built sixty KC-97Es at the Renton plant. The one hundred fifty nine KC-97Fs built had improved R-4360-59B engines rated at 3500 hp.

The KC-97G added 700 gallon underwing fuel tanks. There were five hundred ninety two KC-97Gs built. In 1964 a number of KC-97Gs had J47 jet engines mounted under the outer wing replacing the underwing fuel tank.under the designation KC 97L. Addition of the jet engines increased the speed, which made refueling the faster jet aircraft like the B-58 and F-4 a little easier.

The most important innovation making the KC-97 series so versatile and consequently so useful was that the inflight refueling equipment, including the Flying Boom apparatus, could

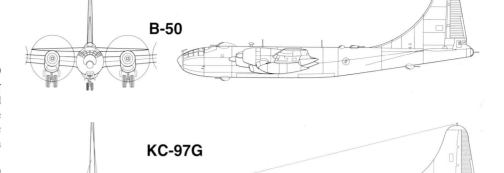

B-50

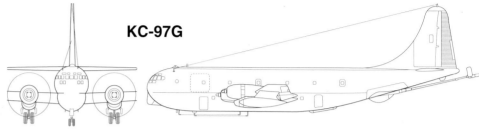

KC-97G

be removed and the cargo doors re-installed for use as a standard transport aircraft. Many C-97s were converted for other mission types including HC-97 search and rescue aircraft, VC-97 flying command posts, and C-97K passenger transports.

Production of the C-97 transport based on the B-50 was 888 aircraft, several times the number of B-50s produced.

The third XC-97 prototype at Muroc Dry Lake AAB in early 1945 with the large cargo doors under the rear fuselage standing open. The XC-97 was based on the B-29 Superfortress utilizing B-29 wings, tail and engines. (David Menard)

A Military Air Transport Command C-97A from the Pacific Division over Pearl Harbor in the early 1950s. The C-97A/KC-97A transport/tanker series was based on the taller tail assembly, P&W R-4360 engines and wings of the B-50. Boeing built a total of 888 C-97/KC-97 aircraft. (Nick William)